Alien Harvester's

TAMSIN O'SHEA

Copyright © 2017 Tamsin O'Shea

All rights reserved.

ISBN:1979286825
ISBN-13:978-1979286824

I wrote this book to help children learn maths with a bit of fun

along the way.

In this one we concentrate on **Circumference of a Circle**.

<u>There is 5 sections to this book.</u>

Section 1: A story, which tells you how find the circumference of a circle.

Section 2: Some guided exercises with pictures.

Section 3: Exercises with pictures.

Section 4: Exercises without pictures.

Section 5: Puzzle pages.

TAMSIN O'SHEA

CONTENTS

	Acknowledgments	i
1	STORY	PG 1-14
2	REFRESHER	15-20
3	PIC GUIDED EXERCISES	21-42
4	PICTURE EXERCISES	43-58
5	STANDARD EXERCISES	59-72
6	PUZZLE PAGES	73-78
7	ANSWERS	79-95

Thankyou for purchasing this Ebook, If you would like a PDF version for printing, feel free to message me at the following with screenshot for proof of purchase. Search for Tamsin O'Shea (author) on Facebook.

I hope you enjoy reading this book, as much as I enjoyed creating it.

Thankyou for your purchase.

x

STORY

"How much longer, until we arrive at our destination?" The captain bellowed.

Everyone froze with fear, the last alien that gave an unsatisfactory answer, was zapped into oblivion. When the captain turned his back, the communications officer pushed his young trainee forward.

The captain spun around and pointed to the young trainee. "You, answer my question."

The poor trainee stammered "W. W. We."

"Out with it, I have better things to do, than stand around here, waiting." The captain said, reaching for his zap gun.

"P. P. Please, captain, they say we will arrive in two zions (days)" the trainee stood there holding his breath, hoping it was satisfactory.

"Very well, I will be in my rooms. Ensure someone informs me, when we are closer to our destination." And the captain left.

"Phew," said the communications officer.

"What do you mean, phew, sir? It was my hide that nearly got zapped. No thanks to you." He grumbled.

"Well, someone had to do it, and you're the newest." Laughed the officer.

Two zions later, they reach their destination. Unfortunately for the poor trainee, he was the one sent, to inform the captain.

"Umm, sir. Captain" the trainee said, gently knocking on the door.

The captain opened the door, zap gun in hand "What!"

The poor trainee, was petrified "So sorry, captain, but we are at our destination, sir."

"Very well." The captain said putting his gun away, and going to the command centre.

With a sigh of relief, the trainee followed the captain, making sure he lagged behind, just enough to run away, if needed.

The captain took his seat, "Are we in position yet?"

"Yes, captain, awaiting your orders." The control alien said.

"We need to get plenty of grain on this trip, prepare the crop circle machine."

"Yes, captain" was the reply.

The alien at the controls, got the suction beam fired up and inputted all the information needed. There was however some measurements that were missing.

"Captain, sir, we need the correct circumference measurement to put into the machine. I'm afraid it wont work without it."

"You should have that already. You buffoon ." the captain growled.

"I'm so sorry, captain, all we have is something called a radius, whatever that is." Bumbled the operator.

"Find a way to work out what we need or you will be visiting oblivion." The captain yelled, then stormed off.

"What do we do now?" Asked the trainee.

The operator scratched his antennae for a minute, before saying "I've got it, we will abduct a human that knows the way of mathematics. Now hop to it."

The trainee found a human called Anna.

The trainee walked into the control center along with Anna, the captain walked in directly behind them "What is this human doing here."

"I would like to know that myself, thank you very much." Anna asked.

"Well, the control operator said we needed a human mathematician, and the sign on her door said that is what she is." The trainee said.

"Get on with it then," the captain growled, "and ensure you wipe her memory, before letting her go. We can't have the humans knowing about us,"

"Yes sir, captain sir."

The control operator pulled Anna over to his work area.

"We need to know how to find the circumference of a circle. Can you help us." He asked.

"Now that I can do. We start with the diameter of the circle, and times it by pi….." Anna started to say.

"We need to know about the circumference human, not how to multiply your earth pies." The captain said, he added with a grumble "Honestly, you humans and your pies, that's all you think about."

Anna shook her head "No, The pi I'm talking about is maths term, not the food one."

"Go on then." The captain said, sitting in his chair.

"Ok, we need to times the diameter of the circle by pi. Pi is 3.141593, it can go on longer but with the calculator on my phone, it automatically adds them in for me." Anna said, pulling out her phone. "Right what is the diameter?"

"Umm, we don't seem to have that. We have something called a radius, will that help?" the control operator asked Anna.

"Yes we can use that. Using the radius we can multiply the radius by 2 x Pi. If you had the diameter, then it would be diameter x Pi" Anna told them.

"So we can go are going to use the radius x 2 = ?. Is that right?" The trainee asked, grinning with pride."

"Yes that's right, so what's the radius measurement that you have?"

"It say's 130 metres." The control operator answered.

"Ok, so 130 x 2 x Pi = 816.81 after we round it to two decimal places." Anna said.

"Well, which number is it then?" the captain roared. He was clearly becoming annoyed.

"Captain, 816.81 is the number you need to input the machine." Anna said.

"Well you could've saved us time, and all that hot air, by simply telling us that. Now, put the details in, so we can get this grain and leave."

Anna asked the control operator, using a quiet voice "What does he mean by get the grain?"

"Well, we often come down and take some grain, for our food. We do this by beaming up a whole circles worth at a time. I think you humans refer to them as being 'crop circles'. Apparently some humans even think it's our way of communicating. It's quite funny really."

Anna stood there rather shocked "Oh."

The operator went ahead and put in the necessary numbers, and set the machine to gather the grain.

"This will take a few zimi's." He said.

"What are zimi's?" Anna asked.

"Oh sorry, in your terms, it's minutes. We use zions as well. Days in human terms." The operator told her.

"Now that you have what you needed, what happens to me? I'm not a fan of being probed you know. Plus I do have a life on earth." Anna asked, with a slight worrying look on her face.

The captain overheard this comment and said, "What is with you humans and probing? You always say the same thing. Seriously, it's as if you want us to, or something."

"Oh, sorry. It's a common thing on our planet. It seems to be the standard alien abduction story." Anna told the captain.

"Well, when we are finished with the grain, we will return you to the earth destination, where you were taken from, and you can let your authorities know, we do not probe, and we take serious offense, to the referral." The captain said, then stormed off.

The trainee led Anna back to the teleportation pods "I'm sorry about the captain, he can be grumpy sometimes. But on a positive note he didn't obliterate you with his raygun."

Anna didn't know if she should be shocked or relieved after that comment. She hopped into the tube, and the trainee sent her back to her office.

2
REFRESHER

Just in case you need a quick refresher.

Starting number is radius

Radius = 5

R x 2 x π = ?

Step 1   Multiply (x , times) the radius by 2. 5 x 2 = 10

Step 2   Using the number you got from Step 1 you multiply it by pi. The symbol on your calculator looks like this π   10 x π = 31.4159

Step 3   You need to round it down to 2 decimal places(because the number is 59 we round up, if it was under 50 then it would be round down. 31.42

So the sum would look like this 10 x π = 31.42

Starting number is <u>diameter</u>

Diameter = 8

D × π = ?

Step 1   Multiply (x , times) the diameter by π 8 × π = 25.13274

Step 2   You need to round it down to 2 decimal places(because the number is 274 we round down, if it was over 500 then it would be round up. 25.13

So the sum would look like this 8 × π = 25.13

Reminder

When rounding up to 2 decimal places: if the number after the 2 decimal places is above 5, 50, 500 etc then it rounds up

Eg 1 .... 6.546 would become 6.55 because the 6 is above 5.

Eg 2 .... 1.2351 would become 1.24 because the 51 is above 50.

Eg 3 .... 3.72849 would become 3.73 because 849 is above 500.

When rounding down to 2 decimal places: if the number after the 2 decimal places is below 5, 50, 500 etch then it rounds down.

Eg1 .... 3.124 would become 3.12 because the 4 is below 5.

Eg 2 ..... 8.2737 would become 8.27 because the 37 is below 50.

Eg 3 .... 1.51221 would become 1.51 because the 221 is below 500.

If the number after the 2 decimal places is 5, 50, or 500 etc then it can be rounded either way you like.

Eg 3.155 could become 3.15 or 3.16 and both would be marked correct.

3

EXERCISES
PICTURE GUIDED

Name_____ Date _____

Exercise 1. This exercise has blank spaces like this _____ which you will need to fill in.

We will start with using Radius for the following exercises.

Radius = 3 cm

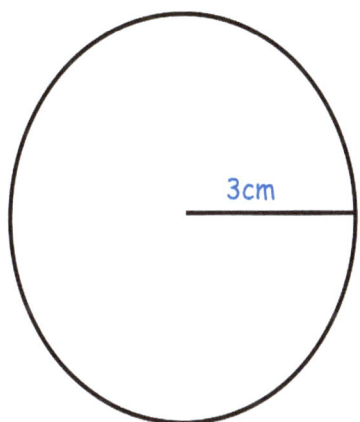

So we multiply the number 3 by 2.   3 x 2 = _____

We then use that number and multiply it by π   6 x π = 18.84956

We need to round it to the nearest 2 decimal points. 18.85 cm

So the sum will look like this 3 x 2 x π = 18.85 cm

The final answer is _____ cm

Name_____ Date _____

Exercise 2. Dont forget to fill in the blanks _____

Radius = 7 m

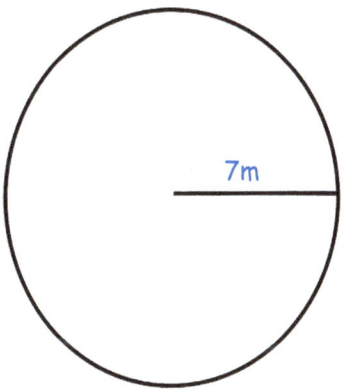

So we multiply the number 7 by 2. 7 x 2 = _____

We then use that number and multiply it by π   14 x π = 43.9823

We need to round it to the nearest 2 decimal points. _____ m

So the sum will look like this 7 x 2 x π = _____ m

The final answer is _____ m

Name_____ Date _____

Exercise 3. Dont forget to fill in the blanks _____

Radius = 12 mm

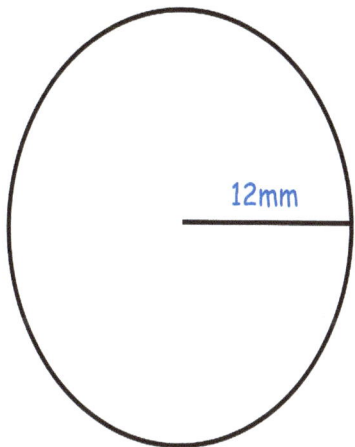

So we multiply the number 12 by 2.   12 x 2 = _____

We then use that number and multiply it by π   _____ x π = 75.3982

We need to round it to the nearest 2 decimal points. _____ mm

So the sum will look like this 12 x 2 x π = _____ mm

The final answer is _____ mm

Name_____ Date _____

Exercise 4. **Dont forget to fill in the blanks** _____

Radius = 2 mm

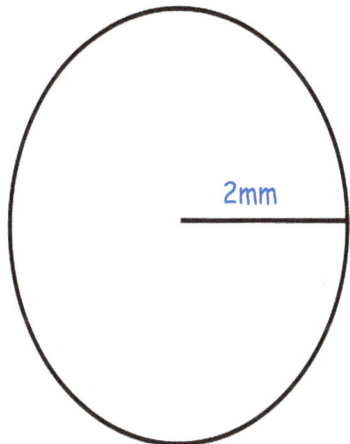

So we multiply the number 2 by 2.   2 x 2 = _____

We then use that number and multiply it by π   _____ x π = _____

We need to round it to the nearest 2 decimal points. _____ mm

So the sum will look like this 2 x 2 x π = _____ mm

The final answer is _____ mm

Name_____ Date _____

Exercise 5. Dont forget to fill in the blanks _____

Radius = 15 m

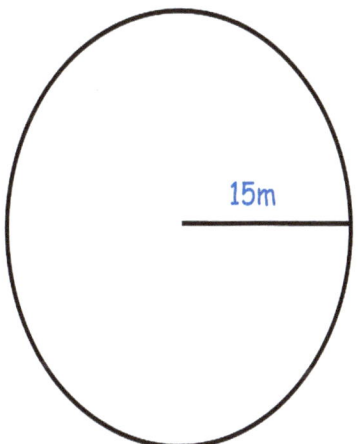

So we multiply the number 15 by 2.   15 x 2 = _____

We then use that number and multiply it by π   _____ x π = _____

We need to round it to the nearest 2 decimal points. _____ m

So the sum will look like this 15 x 2 x π = _____ m

The final answer is _____ m

Name_____ Date _____

Exercise 6. This exercise has blank spaces like this _____ which you will need to fill in.

Now we will use Diameter for the following exercises.

Diameter = 8 cm

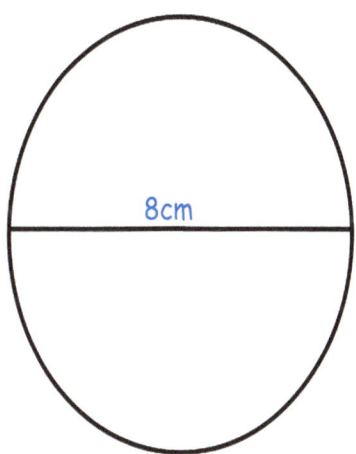

So we multiply the number 8 by π   8 x π = 25.13274

We need to round it to the nearest 2 decimal points. 25._____ cm

So the sum will look like this   8 x π = _____ cm

The final answer is _____ cm

Name_____ Date _____

Exercise 7. Dont forget to fill in the blanks _____

Diameter = 17 mm

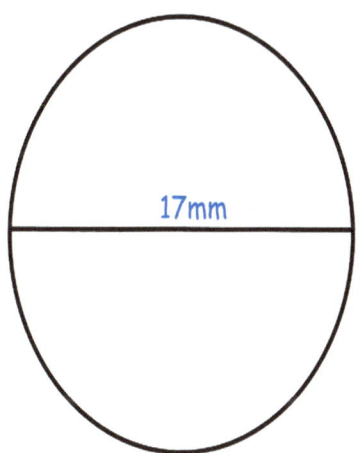

So we multiply the number 17 by π   17 x π = 53.4071

We need to round it to the nearest 2 decimal points. _____ mm

So the sum will look like this   17 x π = _____ mm

The final answer is _____ mm

Name_____ Date _____

Exercise 8. Dont forget to fill in the blanks _____

Diameter = 11 km

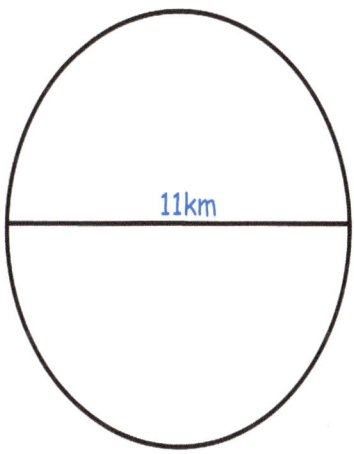

So we multiply the number 11 by π   11 x π = 234.5575

We need to round it to the nearest 2 decimal points. _____ km

So the sum will look like this _____ x π = _____ km

The final answer is _____ km

Name_____ Date _____

Exercise 9. Dont forget to fill in the blanks _____

Diameter = 5 m

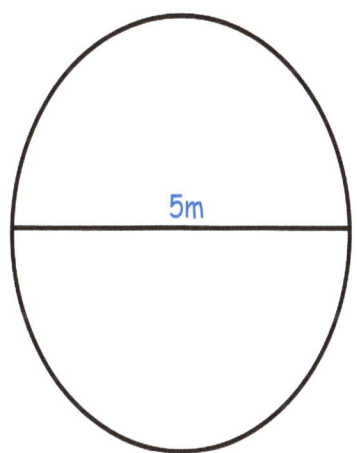

So we multiply the number 5 by π   5 x π = 15.70796

We need to round it to the nearest 2 decimal points. _____ m

So the sum will look like this   _____ x π = _____ m

The final answer is _____ m

Name_____ Date _____

Exercise 10. Dont forget to fill in the blanks _____

Diameter = 19 cm

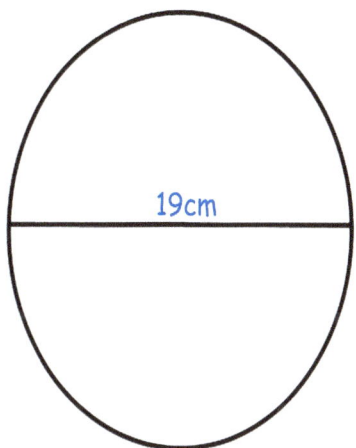

So we multiply the number 19 by π   19 x π = _____

We need to round it to the nearest 2 decimal points. _____ cm

So the sum will look like this _____ x π = _____ cm

The final answer is _____

4

EXERCISES WITH PICTURES

Name_____ Date _____

Exercise 11. Now try these using Radius. **Dont forget to fill in the blanks** _____

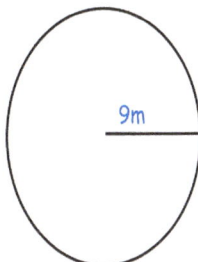
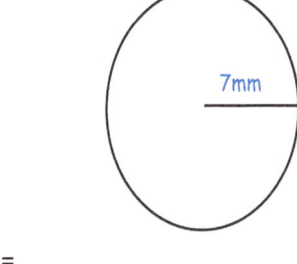

a.   9 x 2 = _____

_____ x π = _____

9 x 2 x π = _____ m

b.   7 x 2 = _____

_____ x π = _____

2 x π = _____ mm

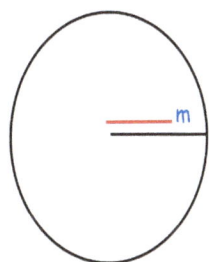
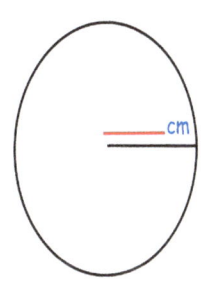

c.   21 x 2 = _____

_____ x π = _____

21 x 2 x π = _____ m

d.   17 x 2 = _____

_____ x π = _____

17 x 2 x π = _____ cm

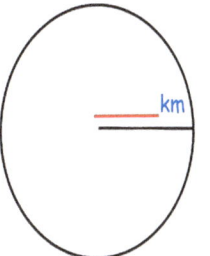
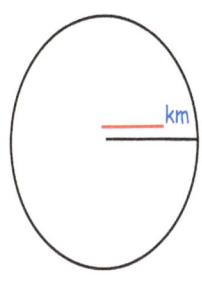

e.   2 x 2 = _____

_____ x π = _____

2 x 2 x π = _____ km

f.   13 x 2 = _____

_____ x π = _____

13 x 2 x π = _____ km

Name_____ Date _____

Exercise 12. Now try these using Radius. **Dont forget to fill in the blanks** _____

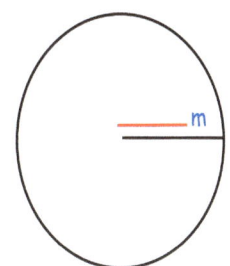　　　　　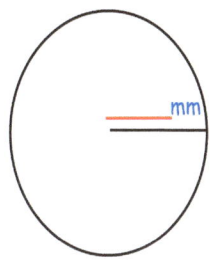

a.　18 x 2 = _____　　　　　b.　22 x 2 = _____

　　_____ x π = _____　　　　　　_____ x π = _____

　　18 x 2 x π = _____ m　　　　　22 x 2 x π = _____ mm

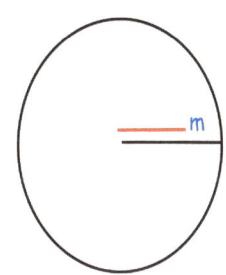　　　　　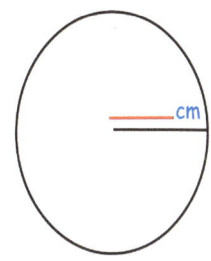

c.　27 x 2 = _____　　　　　d.　3 x 2 = _____

　　_____ x π = _____　　　　　　_____ x π = _____

　　_____ x 2 x π = _____ m　　　　_____ x 2 x π = _____ cm

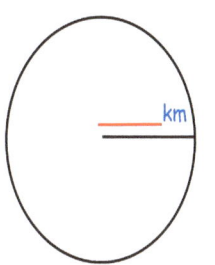　　　　　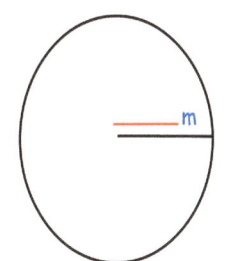

e.　32 x 2 = _____　　　　　f.　16 x 2 = _____

　　_____ x π = _____　　　　　　_____ x π = _____

　　_____ x 2 x π = _____ km　　　_____ x 2 x π = _____ km

Name_____ Date _____

Exercise 13. Now try these using Radius. **Dont forget to fill in the blanks** _____

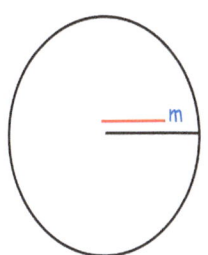

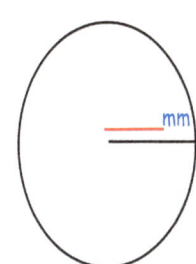

a. 11 x 2 = _____   b. 29 x 2 = _____

   _____ x π = _____   _____ x π = _____

   11 x 2 x π = _____ m   29 x 2 x π = _____ mm

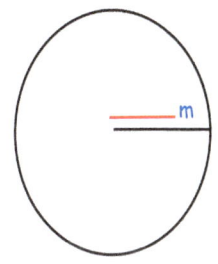

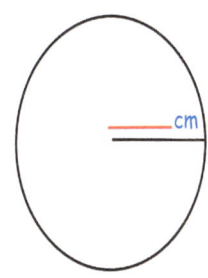

c. 37 x 2 = _____   d. 23 x 2 = _____

   _____ x π = _____   _____ x π = _____

   _____ x 2 x π = _____ m   _____ x 2 x π = _____ cm

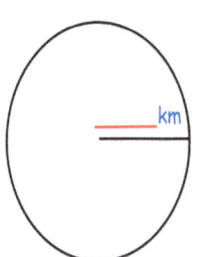

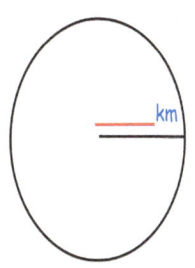

e. 42 x 2 = _____   f. 31 x 2 = _____

   _____ x π = _____   _____ x π = _____

   _____ x 2 x π = _____ km   _____ x 2 x π = _____ km

Name_____ Date _____

Exercise 14. Now try these using Diameter. **Dont forget to fill in the blanks** _____

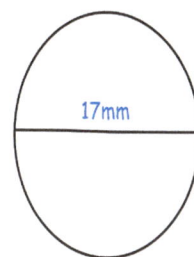 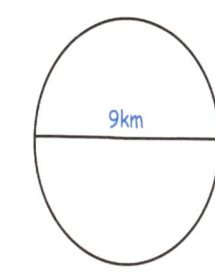

a.   17 x π = _____ mm   b.   9 x π = _____ km

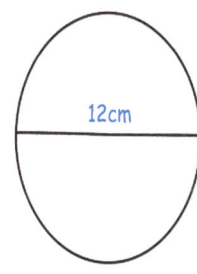 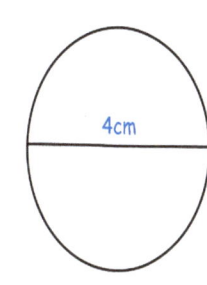

c.   12 x π = _____ cm   d.   4 x π = _____ cm

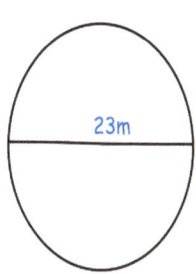 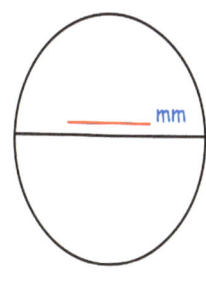

e.   23 x π = _____ m   f.   16 x π = _____ mm

Name_____ Date _____

Exercise 15. Now try these using Diameter. **Dont forget to fill in the blanks** _____

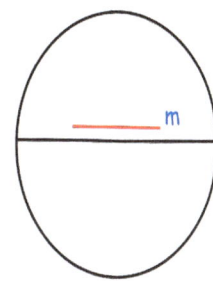
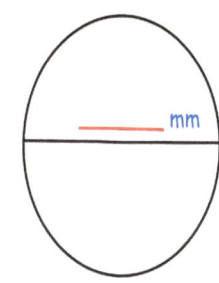

a.   10 x π = _____ m

b.   15 x π = _____ mm

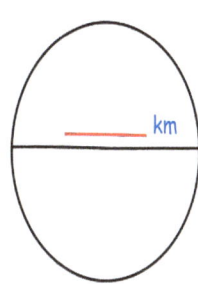
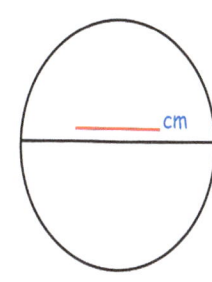

27 x π = _____ km

d.   19 x π = _____ cm

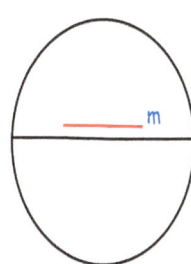
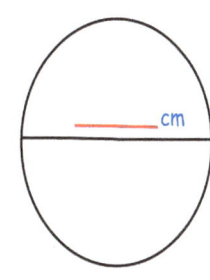

e.   21 x π = _____ m

f.   31 x π = _____ cm

Name_____ Date _____

Exercise 16. Now try these using Diameter. **Dont forget to fill in the blanks** _____

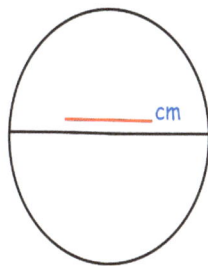
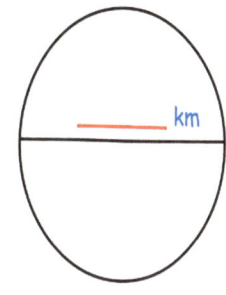

a. 13 x π = _____ cm   b. 7 x π = _____ km

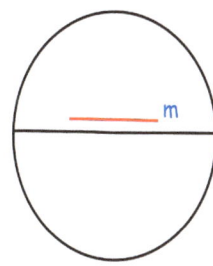
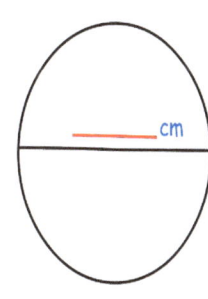

c. 18 x π = _____ m   d. 24 x π = _____ cm

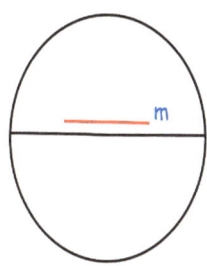
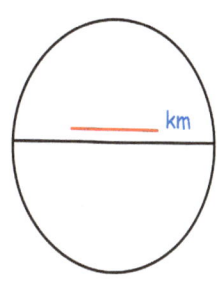

e. 26 x π = _____ m   f. 33 x π = _____ km

Name_____ Date _____

Exercise 17. Now try these using Diameter. **Dont forget to fill in the blanks** _____

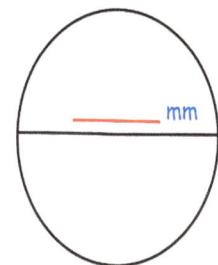
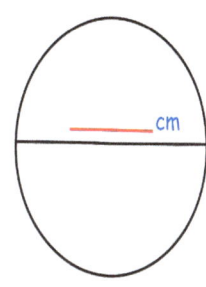

a. 6 x π = _____ mm   b. 20 x π = _____ cm

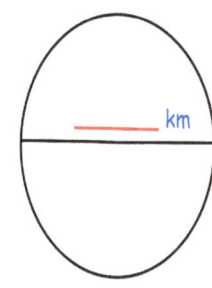
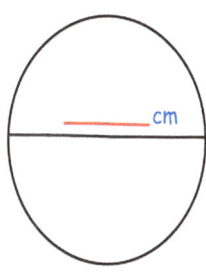

c. 26 x π = _____ km   d. 14 x π = _____ cm

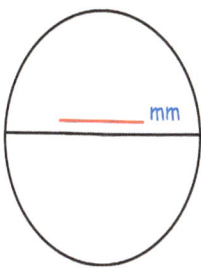
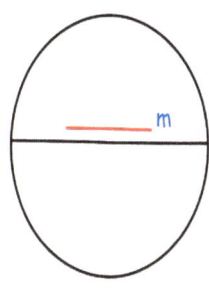

e. 22 x π = _____ mm   f. 34 x π = _____ m

5

EXERCISES NO PICTURES

Name_____ Date _____

Exercise 18. Now try these using Radius. (Plus fill in the Diameter sections The first one is done for you)

Dont forget to fill in the blanks _____

The Exercises from now on will have no pictures.

a. 10 x 2 = _____

   20 x π = _____

   _____ x 2 x π = _____ cm

b. 44 x 2 = _____

   _____ x π = _____

   _____ x 2 x π = _____ mm

c. 19 x 2 = _____

   _____ x π = _____

   _____ x 2 x π = _____ km

d. 26 x 2 = _____

   _____ x π = _____

   _____ x 2 x π = _____ cm

e. 39 x 2 = _____

   _____ x π = _____

   _____ x 2 x π = _____ km

f. 47 x 2 = _____

   _____ x π = _____

   _____ x 2 x π = _____ mm

g. 40 x 2 = _____

   _____ x π = _____

   _____ x 2 x π = _____ m

h. 30 x 2 = _____

   _____ x π = _____

   _____ x 2 x π = _____ km

Name_____ Date _____

Exercise 19. These use Radius. **Dont forget to fill in the blanks** _____

a. 45 x 2 = _____

_____ x π = _____

_____ x 2 x π = _____ cm

b. 28 x 2 = _____

_____ x π = _____

_____ x 2 x π = _____ m

c. 32 x 2 = _____

_____ x π = _____

_____ x 2 x π = _____ m

d. 24 x 2 = _____

_____ x π = _____

_____ x 2 x π = _____ cm

e. 49 x 2 = _____

_____ x π = _____

_____ x 2 x π = _____ mm

f. 36 x 2 = _____

_____ x π = _____

_____ x 2 x π = _____ km

g. 43 x 2 = _____

_____ x π = _____

_____ x 2 x π = _____ km

h. 50 x 2 = _____

_____ x π = _____

_____ x 2 x π = _____ m

Name_____ Date _____

Exercise 20. These use Radius. **Dont forget to fill in the blanks** _____

a. 6 x 2 = _____

   _____ x π = _____

   _____ x 2 x π = _____ mm

b. 25 x 2 = _____

   _____ x π = _____

   _____ x 2 x π = _____ mm

c. 16 x 2 = _____

   _____ x π = _____

   _____ x 2 x π = _____ m

d. 33 x 2 = _____

   _____ x π = _____

   _____ x 2 x π = _____ km

e. 14 x 2 = _____

   _____ x π = _____

   _____ x 2 x π = _____ cm

f. 35 x 2 = _____

   _____ x π = _____

   _____ x 2 x π = _____ m

g. 48 x 2 = _____

   _____ x π = _____

   _____ x 2 x π = _____ km

h. 38 x 2 = _____

   _____ x π = _____

   _____ x 2 x π = _____ m

Name_____ Date _____

Exercise 21. Now try these using Diameter. **Dont forget to fill in the blanks** _____
There is space between sums for working out, if required.

a. 16 x π = _____ m

b. 30 x π = _____ km

c. 26 x π = _____ mm

d. 35 x π = _____ cm

e. 40 x π = _____ mm

f. 49 x π = _____ cm

g. 28 x π = _____ m

h. 34 x π = _____ km

Name_____ Date _____

Exercise 22. Now try these using Diameter. **Dont forget to fill in the blanks** _____
There is space between sums for working out, if required.

a. 22 x π = _____ mm

b. 29 x π = _____ cm

c. 36 x π = _____ mm

d. 13 x π = _____ km

e. 47 x π = _____ cm

f. 38 x π = _____ mm

g. 11 x π = _____ km

h. 32 x π = _____ m

Name_____ Date _____

Exercise 23. Now try these using Diameter. **Dont forget to fill in the blanks** _____
There is space between sums for working out, if required.

a. 46 x π = _____ mm

b. 50 x π = _____ m

c. 41 x π = _____ cm

d. 25 x π = _____ cm

e. 29 x π = _____ mm

f. 44 x π = _____ m

g. 33 x π = _____ km

h. 48 x π = _____ mm

6

PUZZLE PAGES

Name_____ Date _____

Exercise 24. Puzzle page. Using Diameter. **Dont forget to fill in the blanks** _____

There is a letter under each of the sums. Use the answer to fill in the required letter below, to solve the puzzle. The first one , the letter D, is done for you.

a. 3 x π = 9.42 m
  D

b. 25 x π = _____ mm
  L

c. 2 x π = _____ km
  W

d. 13 x π = _____ cm
  O

e. 18 x π = _____ mm
  N

f. 14 x π = _____ cm
  E

6.28	43.98	78.54	78.54

9.42	40.84	56.55	43.98
D			

Name_____ Date _____

Exercise 25. Puzzle page. Using Diameter. **Dont forget to fill in the blanks _____**

There is a letter under each of the sums. Use the answer to fill in the required letter below, to solve the puzzle. The first one, the letter R, is done for you.

a. 21 x π = 65.97 mm
   R

b. 34 x π = _____ cm
   K

c. 28 x π = _____ km
   T

d. 39 x π = _____ m
   O

e. 49 x π = _____ mm
   W

f. 37 x π = _____ mm
   E

g. 26 x π = _____ m
   G

h. 43 x π = _____ cm
   A

1.68	65.97	116.24	135.09	87.96
	R			

153.94	122.52	65.97	106.81

7
ANSWERS

Exercise 1.

So we multiply the number 3 by 2.   3 x 2 = __6__

We then use that number and multiply it by π   6 x π = 18.84956

We need to round it to the nearest 2 decimal points. 18.85 cm

So the sum will look like this   3 x 2 x π = 18.85 cm

The final answer is __18.85__ cm

Exercise 2.

So we multiply the number 7 by 2.   7 x 2 = __14__

We then use that number and multiply it by π   14 x π = 43.9823

We need to round it to the nearest 2 decimal points. __43.98__ m

So the sum will look like this   7 x 2 x π = __43.98__ m

The final answer is __43.98__ m

Exercise 3.

So we multiply the number 12 by 2.   12 x 2 = __24__

We then use that number and multiply it by π   __24__ x π = 75.3982

We need to round it to the nearest 2 decimal points. __75.40__ mm

So the sum will look like this   12 x 2 x π = __75.40__ mm

The final answer is __75.40__ mm

Exercise 4.

So we multiply the number 2 by 2. $2 \times 2 =$ __4__

We then use that number and multiply it by π __4__ $\times \pi =$ __12.56637__

We need to round it to the nearest 2 decimal points. __12.57__ mm

So the sum will look like this $2 \times 2 \times \pi =$ __12.57__ mm

The final answer is __12.57__ mm

Exercise 5.

So we multiply the number 15 by 2. $15 \times 2 =$ __30__

We then use that number and multiply it by π __30__ $\times \pi =$ __94.2478__

We need to round it to the nearest 2 decimal points. __94.25__ m

So the sum will look like this $15 \times 2 \times \pi =$ __94.25__ m

The final answer is __94.25__ m

Exercise 6.

So we multiply the number 8 by π $8 \times \pi =$ 25.13274

We need to round it to the nearest 2 decimal points. 25.__13__ cm

So the sum will look like this $8 \times \pi =$ __25.13__ cm

The final answer is __25.13__ cm

Exercise 7.

So we multiply the number 17 by π   17 x π = 53.4071

We need to round it to the nearest 2 decimal points. __53.41__ mm

So the sum will look like this   17 x π = __53.41__ mm

The final answer is __53.41__ mm

Exercise 8.

So we multiply the number 11 by π   11 x π = 34.5575

We need to round it to the nearest 2 decimal points. __34.56__ km

So the sum will look like this   __11__ x π = __34.56__ km

The final answer is __35.56__ km

Exercise 9.

So we multiply the number 5 by π   5 x π = 15.70796

We need to round it to the nearest 2 decimal points. __15.71__ m

So the sum will look like this   __5__ x π = __15.71__ m

The final answer is __15.71__ m

Exercise 10.

So we multiply the number 19 by π   19 x π = __59.6903__

We need to round it to the nearest 2 decimal points. __59.69__ cm

So the sum will look like this   __19__ x π = __59.69__ cm

The final answer is __59.69__ cm

Exercise 11.

a. 9 x 2 = __18__
   __18__ x π = __56.5487__
   9 x 2 x π = __56.55__ m

b. 7 x 2 = __14__
   __14__ x π = __43.9823__
   7 x 2 x π = __43.98__ mm

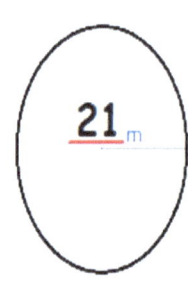

c. 21 x 2 = __42__
   __42__ x π = __131.947__
   21 x 2 x π = __131.95__ m

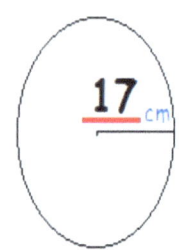

d. 17 x 2 = __34__
   __34__ x π = __106.8142__
   17 x 2 x π = __106.81__ cm

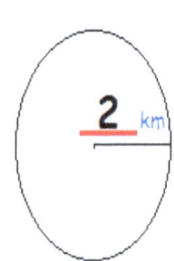

e. 2 x 2 = __4__
   __4__ x π = __12.56637__
   2 x 2 x π = __12.57__ km

f. 13 x 2 = __26__
   __26__ x π = __81.6814__
   13 x 2 x π = __81.68__ km

Exercise 12.

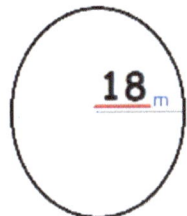
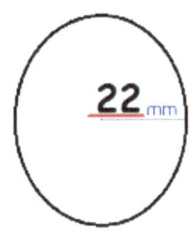

a. 18 x 2 = __36__
    __36__ x π = __113.0973__
    18 x 2 x π = __113.10__ m

b. 22 x 2 = __44__
    __44__ x π = __69.115__
    22 x 2 x π = __69.11__ mm
    or 69.12

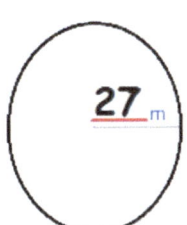
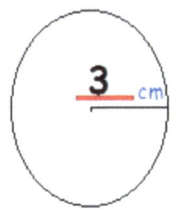

c. 27 x 2 = __54__
    __54__ x π = __169.646__
    __27__ x 2 x π = __169.65__ m

d. 3 x 2 = __6__
    __6__ x π = __18.84956__
    __3__ x 2 x π = __18.85__ cm

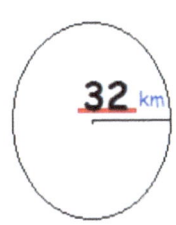
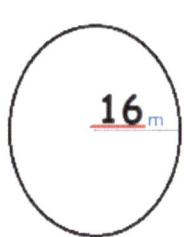

e. 32 x 2 = __64__
    __64__ x π = __201.062__
    __32__ x 2 x π = __201.06__ km

f. 16 x 2 = __32__
    __32__ x π = __100.531__
    __16__ x 2 x π = __100.53__ km

Exercise 13.

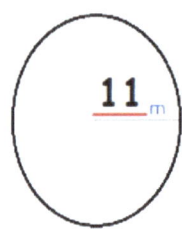

a. 11 x 2 = **22**
    22 x π = **69.115**
    11 x 2 x π = **69.11** m
    or **69.12**

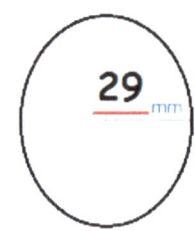

b. 29 x 2 = **58**
    58 x π = **182.2124**
    29 x 2 x π = **182.21** mm

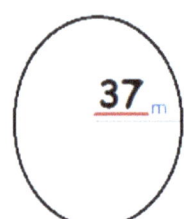

c. 37 x 2 = **74**
    74 x π = **232.478**
    37 x 2 x π = **232.48** m

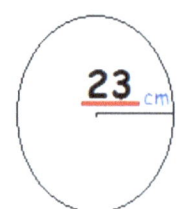

d. 23 x 2 = **46**
    46 x π = **144.5133**
    23 x 2 x π = **144.51** cm

e. 42 x 2 = **84**
    84 x π = **263.894**
    42 x 2 x π = **263.89** km

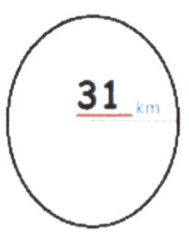

f. 31 x 2 = **62**
    62 x π = **194.7787**
    31 x 2 x π = **194.78** km

Exercise 14.

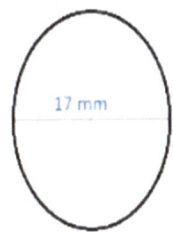
a. 17 × π = **53.41** mm

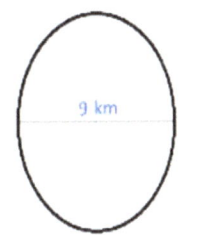
b. 9 × π = **28.27** km

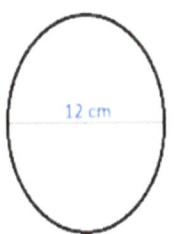
c. 12 × π = **37.70** cm

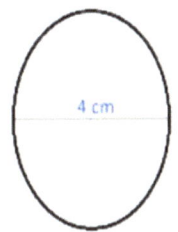
d. 4 × π = **12.57** cm

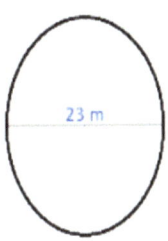
e. 23 × π = **72.26** m

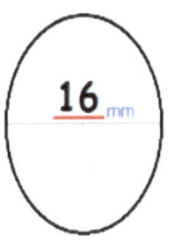
f. 16 × π = **50.27** mm

Exercise 15.

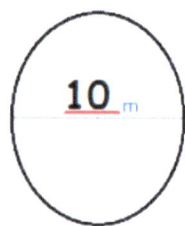
a. 10 × π = **31.42** m

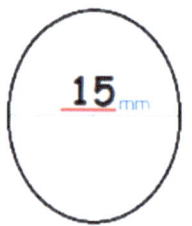
b. 15 × π = **47.12** mm

c. 27 × π = **84.82** km

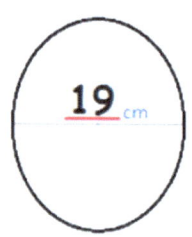
d. 19 × π = **59.69** cm

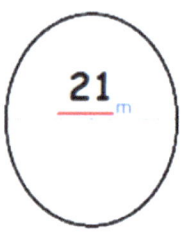
e. 21 × π = **65.97** m

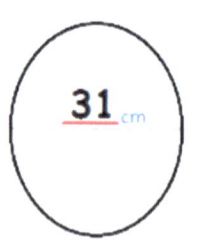
f. 31 × π = **97.39** cm

Exercise 16.

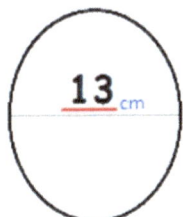
a. 13 × π = **40.84** cm

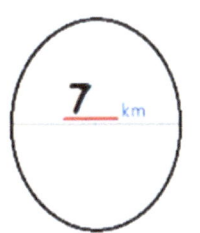
b. 7 × π = **29.99** km

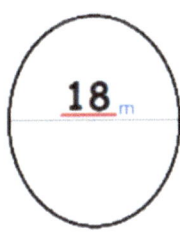
c. 18 × π = **56.55** m

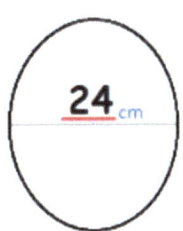
d. 24 × π = **75.40** cm

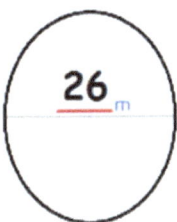
e. 26 × π = **81.68** m

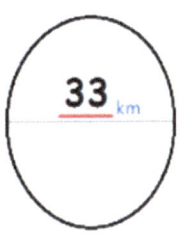
f. 33 × π = **103.67** km

Exercise 17

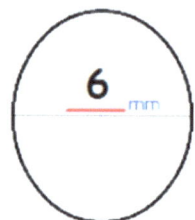
a. 6 × π = **18.85** mm

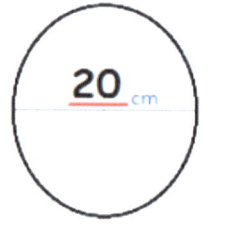
b. 20 × π = **62.83** cm

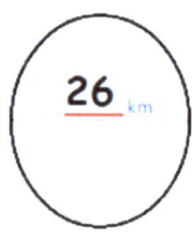
c. 26 × π = **81.68** km

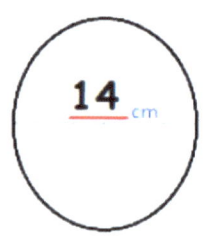
d. 14 × π = **43.98** cm

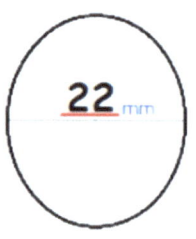
e. 22 × π = **69.11** mm
or **69.12**

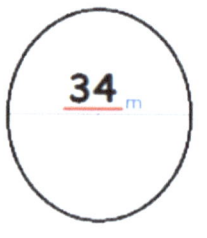
f. 34 × π = **106.81** m

Exercise 18.

a. 10 x 2 = **20**
   20 x π = **62.8319**
   10 x 2 x π = **62.83** cm

b. 44 x 2 = **88**
   88 x π = **276.46**
   44 x 2 x π = **276.46** mm

c. 19 x 2 = **38**
   38 x π = **119.3805**
   19 x 2 x π = **119.38** km

d. 26 x 2 = **52**
   52 x π = **163.3628**
   26 x 2 x π = **163.36** cm

e. 39 x 2 = **78**
   78 x π = **245.044**
   39 x 2 x π = **245.04** km

f. 47 x 2 = **94**
   94 x π = **295.31**
   47 x 2 x π = **295.31** mm

g. 40 x 2 = **80**
   80 x π = **251.3274**
   40 x 2 x π = **251.33** m

h. 30 x 2 = **60**
   60 x π = **188.4956**
   30 x 2 x π = **188.50** km

Exercise 19.

a. 45 x 2 = __90__
   __90__ x π = __282.7433__
   __45__ x 2 x π = __282.74__ cm

b. 28 x 2 = __56__
   __56__ x π = __175.9292__
   __28__ x 2 x π = __175.93__ m

c. 32 x 2 = __64__
   __64__ x π = __201.062__
   __32__ x 2 x π = __201.06__ m

d. 24 x 2 = __48__
   __48__ x π = __150.7964__
   __24__ x 2 x π = __150.80__ cm

e. 49 x 2 = __98__
   __98__ x π = __307.876__
   __49__ x 2 x π = __307.88__ mm

f. 36 x 2 = __72__
   __72__ x π = __226.1947__
   __36__ x 2 x π = __226.19__ km

g. 43 x 2 = __86__
   __86__ x π = __270.177__
   __43__ x 2 x π = __270.18__ km

h. 50 x 2 = __100__
   __100__ x π = __314.159__
   __50__ x 2 x π = __314.16__ m

Exercise 20.

a. 6 × 2 = __12__
   __12__ × π = 37.6991
   __6__ × 2 × π = __37.70__ mm

b. 25 × 2 = __50__
   __50__ × π = 157.0796
   __25__ × 2 × π = __157.08__ mm

c. 16 × 2 = __32__
   __32__ × π = 100.531
   __16__ × 2 × π = __100.53__ m

d. 33 × 2 = __66__
   __66__ × π = 207.345
   __33__ × 2 × π = __207.34__ km
   or 207.35

e. 14 × 2 = __28__
   __28__ × π = 87.9646
   __14__ × 2 × π = __87.96__ cm

f. 35 × 2 = __70__
   __70__ × π = 219.9115
   __35__ × 2 × π = __219.91__ m

g. 48 × 2 = __96__
   __96__ × π = 301.593
   __48__ × 2 × π = __307.59__ km

h. 38 × 2 = __76__
   __76__ × π = 238.761
   __38__ × 2 × π = __238.76__ km

Exercise	A	B	C	D	E	F	G	H
21	50.27	94.25	81.68	109.96	125.66	153.94	87.96	106.81
22	69.11	90.11	113.1	40.84	147.65	119.38	34.56	100.53
23	14451	157.08	128.81	75.54	91.11	138.23	103.67	150.8

Exercise	A	B	C	D	E	F
24	9.42	78.53	6.28	40.84	56.55	43.98

6.28	43.98	78.54	78.54
W	E	L	L

9.42	40.84	56.55	43.98
D	O	N	E

Exercise	A	B	C	D	E	F	G	H
25	135.09	106.81	87.96	12.52	153.94	116.24	81.68	135.09

81.68	65.97	116.24	135.09	87.96
G	R	E	A	T

153.94	122.52	65.97	106.81
W	O	R	K

ABOUT THE AUTHOR

Tamsin O'Shea is a proud mother and grand-mother. She enjoys reading and writing. Both her children and her grand-children are inspiration for her books.

She has many books near completion, and some already published. They consist of standard children's books, YA Books and Story/Textbooks, where she teaches math using the story.

You can contact via facebook. If you search Tamsin O'Shea Author

www.ingramcontent.com/pod-product-compliance
Lightning Source LLC
Chambersburg PA
CBHW051153220526

45473CB00003B/758